THE ADVENTURES OF
THE RABBI WHO BECOMES A FLIGHT ATTENDANT

By
David Pinkwasser

Copyright © David Pinkwasser 2024

All Rights Reserved

No part of this publication may be reproduced, stored in a retrieval system, or transmitted in any form or by any means, electronic, mechanical, photocopying, recording, or otherwise, without the written permission of the author or the publisher.

Contents

Dedication i
About the Author ii
Preface 1
Introduction 2
Background 6
It Happened in the Cemetery 8
A Spooky Time at the Cemetery 15
Can A Wedding Dress Be Too Short? 18
A Brit Milah (Ritual Circumcision) The Baby's Revenge 22
Don't Throw That At Me! 25
I Thought the Roof Was Coming Down 30
Interfaith Thanksgiving Service Lifelong Friendships 33
I Want A Passover Seder, Too! 45
The Rabbi Leads Services At the Methodist Church 48
I Brought Jewish Special Education to Arizona 51
The Rabbi Faints on the Pulpit 55
Don't Shoot—I Washed My Hands! 59
Teens Polish the Cemetery 62
The Tefillin Police 66
When Singing Is Inappropriate 70
The Mile High Club Is It Open to the Public? 74
Nice Flashlight, Buddy 79
Is That A Flashbulb Or Lightning? 85
I Think I Smell Smoke! Is It A Cigarette Or A Fire? 89
We're Gonna Doy! We're Gonna Doy! (Die in Brooklyn) 92
I Left My Heart in Havana 96
It's Thanksgiving What's That Aroma? 100
Can You Synchronize This? 105
Is There A Doctor In The House? 108
Serious Stuff—CISM-CoHeart-Lodo 116
Help! Help! I'm Stuck!! 125
Funny and Silly Stuff 128
Epilogue 132

Dedication

I'd like to dedicate this book to my wife, Ann and my sons, Josh and Aaron. Without their love and support, I would not have been able to do all the "crazy" things I did in my life. Thank you for being supportive and always being there.

About the Author

David Pinkwasser is a retired Reform Rabbi. He was raised Conservative, educated Orthodox and served Reform Jewish congregations for 25 years. After retiring from the rabbinate, he became a flight attendant for a major airline. You will find some serious and touching moments, as well as hilarious happenings, both as a rabbi and flight attendant.

Preface

This book is totally factual. The names of people mentioned have been changed to protect privacy. You will find deep meaning in what both a Rabbi does, and flight attendants do in an atmosphere of lightness and humor. Some of the stories told will be interesting and touching, and others will make you laugh. Being both a rabbi and a flight attendant blends together the reality of life, which is both serious and laughable. It was my life. I am so pleased to be able to share this with you and hope that it gives you inspiration to follow your dreams.

David Pinkwasser

Introduction

Most of you think that you know what a clergy and a flight attendant do. I am here to tell you that you DON'T! You know what you see, but that is just the tip of the iceberg. I remember the words of my father, commenting on my rabbinic career, who said,

"You have a great job. You get up once or twice a week and make a speech, and then they pay you a bundle."

Unfortunately, that is what many people think.

As a flight attendant, people think that this is a cool job. You serve a drink or a meal and then get to travel all around the world. Not bad!

WRONG!!!!!

Let's start with the clergy. The part you see is the "Pretty Part." You stand as the leader before a house full of people with beautiful music, inspiring words and happy faces. BUT THERE'S MORE!

*THE ADVENTURES OF THE RABBI
WHO BECOMES A FLIGHT ATTENDANT*

1. Life Cycle events (Weddings, Bar/Bat Mitzvot, Brit Milah Ceremonies, Funerals, Unveilings, Conversion Ceremonies etc.)

2. Board obligations (Board of Rabbis, Ministerial Association, Israel Bonds Board, Jewish Federation Board, etc.)

3. Teaching (Adult Education, Hebrew High School, Bar/Bat Mitzvah Class, Torah Study)

4. Counseling (Family, Personal, Religious)

5. Various unscheduled meetings (Hospital visits, Care Center visits, Choir, Youth Group, Religious School, Confirmation Class)

6. Emergencies (Available 24 hours a day for medical, death or personal need)

To put it in perspective:

1. Weddings 75+ a year

2. Bar/Bat Mitzvah 40+ a year

3. Brit Milah (Ritual Circumcisions) 25+ a year

4. Funerals 75+ a year

5. Unveilings (Tombstone Ceremony after 1 year) 30+

6. Conversion Ceremonies 1 or 2 a year (Group Ceremony)

7. Board meetings are monthly

8. Teaching is weekly

9. Unscheduled meetings, as needed

10. In addition—I also visited the Jews in prison monthly and spent several years at the Arizona State Hospital as a Jewish Chaplain

Now For Flight Attendant Duties:

1. Know all of the FAA regulations

2. Medical training

3. Fire fighting and prevention

4. Hijack prevention and self-defense

5. Procedures in case of an emergency or water landing

6. Defusing irate passengers

7. Evacuation procedures

8. Situational awareness

As you can see, it is more than just the "Pretty Part." This book will expose you to the real life of a clergy and a flight attendant. You will not only get the superficial, but you'll see what these careers are really like. Yes, they are cool jobs, but not what you thought. Enjoy!

Background

Those who become members of the clergy are dedicated souls. You don't do this on a whim. The training is grueling, and the time spent learning is intense. You do it because you feel you can make a difference. As mentioned before, the job of a rabbi is multifaceted. You are there to serve your congregation and community. You wear many different hats, or in my case, yarmulkes, and you are there to bring your congregation closer together to each other and to God. Not an easy task. This is where being multi-skilled and super flexible makes it work.

I have been trained in many fields. I have studied over 8 languages and can speak most of them fluently. I was also trained in Special Education to help those who had special needs, and I had a myriad of jobs while in Junior High and High School that prepared me for the challenge. Growing up in New York City, one is capable of having many interesting and non-interesting jobs that gave me perspective as to what I wanted to do and what I did not want to do. It has made my career and life most meaningful and fulfilling. Let's start with a hilarious story.

David Pinkwasser

It Happened in the Cemetery

(A Strange Place to Start—You'll See—**Watch the Rabbi**)

I hear a voice calling: Help! Help me!! Get me out of here!!!

The voice sounded familiar, even though I did not know anyone in this cemetery. The voice sounds familiar—A lot like my own. How weird. Anyway, I ignored it and went one with my duties for the day.

You may think that this should be the end of this book, not the beginning. However, people are very interested in funerals because death is a taboo subject in the United States. People don't want to talk about it until the time that a loved one passes away, and then they run to their clergy or funeral director for guidance. They are in unchartered territory.

When clergy get together, we often talk about funerals and services that we have given to the dying. If you are at a cocktail party and the subject comes up, those who are non-clergy quickly run away. If you want to empty the room, this is how it is done. Keep talking, and no one will be left other than the clergy.

We have made cemeteries more palatable by planting beautiful shade trees and bushes. The earth that is going to be replaced in the grave is often covered with artificial turf to make it less harsh.

The following story is one that takes place at the County Cemetery. I have volunteered to do indigent burials every 3 months. None of these people are Jewish. They are poor people who sometimes have some family and friends accompanying them, and other times, there is no one.

The reason clergy do not find cemeteries depressing is that this is one of the many venues in which we work, just as sanctuaries, hotels, meeting rooms and doctors' offices are our venues for weddings, ritual circumcisions and other lifecycle events.

The situation at the County Cemetery is much different. Very few trees, dirt, and no grass, tombstones that are metal discs about 5 inches in diameter with the names of the deceased and the year they died. That's it!

This morning, I am going to do 8 funerals. The county has dug 20 graves, and 12 are covered with ¼ inch plywood. This closes the grave visually, but if you step on it, you'll fall in. The grave diggers are prisoners from the county jail in their orange jumpsuits. Every half hour, another body is brought in from a funeral home that has contracted with the county. Usually, the caskets are brought in by pickup truck, van, el Camino, and rarely a hearse.

It is my goal to make those coming to this place leave with a little bit of closure and hope. I want to lift their spirits a bit.

THE ADVENTURES OF THE RABBI
WHO BECOMES A FLIGHT ATTENDANT

Keep your eye on the rabbi.

I have now done the first 2, and for number 3, there is no family or mourners. I find out some information from the funeral director and make up a little eulogy. The casket is lowered into the grave. Then, 2 women show up who are cousins of the deceased. They only speak Spanish. They want to know if they can see their cousin one more time. I translated for the funeral director, who obliged and jumped into the grave with his feet propped on the burial vault. He opens the casket, and the deceased is naked. Yes, stark naked! The women gasp and turn their heads away. The funeral director closes the casket and I now proceed to continue the service in Spanish.

I scolded the funeral director. I asked why he did that if he knew that the person was not dressed. The response was that they wanted to see their cousin. Some people have no common sense. What can I say?

The morning gets worse. Remember—**Watch The Rabbi.**

Numbers 4-7 go smoothly without a hitch. I am now on grave 8. I finish the service and am now walking from the front to the back to get some earth to place into the grave. As I walk between 8 and 9 there is a small spot about 4 inches in diameter that is uncovered. The rest of grave 9 is covered. As I walk over this small spot, my right foot falls into 9. Then the ground gives way, and the other footfalls in. I am now in grave 9 with my legs dangling. One hand on the wood and the other grasping the dirt, dressed in my black suit. I am hanging by my armpits. The 2 grave diggers gasp and come to grab me and pull me out. The mourners are all aghast, and they ask, "Are you all right?" I, of course, say that I am fine because I am mortified. I brush myself off, finish the service and go back to my office.

After 30 minutes at my desk, I am all scratched up and bruised, and I feel terrible. I go home.

THE ADVENTURES OF THE RABBI
WHO BECOMES A FLIGHT ATTENDANT

I'm lying on the couch in a pair of gym shorts and shirtless because the shirt was rubbing against my scratches. My wife comes home and wants to know why I am home in the middle of the day. I responded that I was in an accident.

She asked, "With the car?"

"No, it was an industrial accident."

The response was, "No, you didn't. I know where you were."

I groaned, "Yes, I did."

She then proceeded to call her mother, sister and anyone she could think of to tell them what happened. Somehow, I could not share the humor.

One month later, I am presiding at a funeral at a different cemetery, and the funeral director is following me around. He keeps walking behind me and has his hands out as if to catch me if I fall. I glance back and ask, "Charley, what are you doing?"

He responds, "I want to make sure that you're safe."

I then said, "Oh, you heard what happened over at the County Cemetery."

He answered, "O yeh! We all heard about it."

I do need to tell you that this was before cell phones, Wifi, faxes or social media. The event was such a hoot that all the funeral directors in town called each other to get a chuckle on my behalf.

I am happy to say that this has never happened again. Most funerals have solid ground around the gravesite. Also, I was very cautious the next time I went out to the County Cemetery. Sometimes, doing a good deed has a price. I was willing to pay that price to do what I did.

THE ADVENTURES OF THE RABBI
WHO BECOMES A FLIGHT ATTENDANT

A Spooky Time at the Cemetery

The sun was shining, and then the wind began to blow **harder and harder, but only around the graveside.**

As I have mentioned before, cemeteries are not scary places for clergy. We come here often, and our goal is to comfort a family and give them the closure that they need. This time was STRANGE!

I was called by a funeral home to do a service. I did not know the family and got all of the information on the phone. The first time I met them was at the interment.

The weather was beautiful, as it is in Arizona most of the time, and the funeral director and I were chatting. Then came the family. That's when the turbulence began. I thought it was strange, but you can't predict Mother Nature.

I was now ready to place the memorial ribbons on the family, called Kriya Ribbons. I then cut a piece of the ribbon to symbolize that a piece of it has been cut away. Jews are very familiar with this tradition. HOWEVER, one of the family members came with a low cut, strapless dress that was very revealing. It was much more appropriate for a social event than a funeral. The woman was the daughter of the deceased and told me that this was the only black dress that she had. Okay, I can go with the flow, but where do I pin the ribbon? It is usually worn about chest level, much like a soldier would wear their medals.

I decided that the only logical thing was to give her the ribbon and let her pin it on. She placed it at the end of her breast. It was like a Las Vegas stage show, except the showgirl had on only one tassel. I cut the ribbons of the other family members and gave her the razor to cut her own.

The service continued. The dust was blowing and getting into my eyes. It was bothering my contact lens, so I had tears coming out of my eyes. The family was so impressed at how emotional the service was that even I was crying and did not know the deceased.

When the family left, the wind stopped, and the sun shined brightly. The funeral director and I tried to rationalize what we saw. Neither could come up with a logical explanation.

The only thing that I could gather is that the relationship between the family and the deceased was not good, and their presence was not well appreciated. If you don't believe that there is spiritual energy out there, this event could make you a believer.

Many times, when families tell me glowing things about a loved one, it is often embellished or sometimes not true. It is a way they would like to remember them, which often didn't exist, or it was a way to present a loved one to the public.

Funerals are always unique, and you never know what will play out. Flexibility is the key.

NO MORE FUNERALS AND SAD STUFF—WE MOVE ON!!!

David Pinkwasser

Can A Wedding Dress Be Too Short?

Is it polite to make wolf whistles at a wedding when the bride comes down the aisle? Probably not, nor did it happen here, but it could have easily.

I officiate at many wedding ceremonies each year. I am usually not a stickler as to what the bridal party is wearing. Some of my more traditional colleagues make strict rules as to what the couple is wearing, especially the bride. Modesty during the ceremony is a major concern.

You must remember that I was ordained in the early 70's. Bridal dresses were long and flowing with veils and lace, covering the upper part of the bride and trains following behind. The mini dress had become popular, but not for wedding gowns. I was not concerned if there was too much arm or shoulder showing. Today changed it all.

The bride wore a super short mini dress with a slit up the front. The top part plunged down to almost her navel, and there she was. I must say, she was a very beautiful woman. The dress would have been more appropriate for a club or social gathering.

At this wedding, my family was invited. My 2 sons were about 7 and 10. When the bride processed in, they almost fell out of their chairs. Neither had ever been allowed to see that much flesh on a grown woman.

I, as the rabbi, normally focus on the couple, but today, I spent a lot of time looking at my little black book and tried not to be distracted by the spectacle.

After this situation, I came up with a set of rules for brides at the wedding. I'm not as strong as some of my Orthodox colleagues, but at least there would be a sense of spirituality at the ceremony and not the feeling of being at a Vegas show.

Many years have gone by, and now I am counseling a couple before their wedding. I talked about proper dress and told the story of the mini bride's dress. The bride turned red, and the groom began to chuckle. I asked what the reaction was all about. They told me that this was the dress that the bride picked out. I told them that it was inappropriate and there were still several months ahead to find something more fitting.

The groom responded that he liked the way she looked in the dress. She is very pretty, and this dress showed off her beautiful body. My response to him is that it is true that she is very good-looking. Why not get something very sexy and revealing for the honeymoon, where just the 2 of you can enjoy it and not 150 of your guests who are onlookers?

With much sorrow, they took my advice. At the wedding, the bride wore a beautiful peasant dress that was off the shoulders a bit but very tasteful. She was a knockout.

Sometimes, people are so taken back by the wedding experience and the moment that they need a person with a level head to guide them. That's where I come in. I am not overwhelmed and can

THE ADVENTURES OF THE RABBI
WHO BECOMES A FLIGHT ATTENDANT

give some stable advice that they, in the long run, will appreciate.

The work of a clergy is never dull!

David Pinkwasser

A Brit Milah (Ritual Circumcision) The Baby's Revenge

"I'll get him," says the baby. Okay, babies can't talk, but they have other ways of communicating. This one was it. To quote NYC slang, it was a pisser!

I officiate at several dozen Brit Milahs, commonly called a Bris, each year. It is performed either in the person's home, with a reception afterward, or in the doctor's office, generally with only parents and possibly grandparents in attendance.

Bris ceremonies on the weekend were no problem. Our offices were closed, both mine and the doctor. On weekdays we usually scheduled them around 5:00 - 5:30 pm, after both the doctor and I were finished with our usual office hours. I worked with several Jewish pediatricians who were familiar with the ritual and understood its religious implications.

The Bris was commanded in the Torah to be performed when the baby was 8 days old. It was a sign of being part of the covenant between God and the Jewish people. It went back to the time of Abraham. It is the oldest of rituals in the Jewish faith.

Today, the doctor and I decided that it would be nice that after the ceremony, our wives would join us, and we'd go out to dinner. Normally, it is just the 2 of us and the family, no spouses.

Now, we begin. The doctor removed the diaper from the baby, and the child proceeded to send a stream of urine up into the air and onto the doctor. I giggled a bit, but soon, the doctor grabbed the diaper to cover the flood.

I, in the meantime, am on the other side of the baby. I have a glass of sweet wine, which is used in the ceremony after the surgical part, and with a gauze pad, I am letting the baby suck on it to help calm him down. This is something that I always did. NOW, as the doctor is ready to continue, the baby lets go of a giant explosion of poop that gets all over him. I am trying not to laugh, but fortunately, we are both close enough friends that he didn't get offended. At this point, things change, the doctor, now covered in pee and poop, is doing his thing, and THEN the baby starts throwing up all over me. (I guess I shouldn't have laughed so soon!) We finish the procedure the ritual, and now it is time for dinner.

We both decided that it might be better if we headed home, showered and changed our clothing. We met an hour later and had a lovely dinner with the wives and we laughed all throughout the meal.

This has never happened before and has never happened again. You never know where life is going to take you.

THE ADVENTURES OF THE RABBI WHO BECOMES A FLIGHT ATTENDANT

Don't Throw That At Me!

I was dressed in my rabbinic robe, with my holy prayer shawl, the tallit, and being pelted with objects as I stood on the pulpit. It was like an old vaudeville show where people brought fruit and other objects to throw at the entertainers if they didn't like the performance. This one was a crazy day!

As rabbi, I was in charge of preparing boys and girls for their Bar/Bat Mitzvah. The name Bar Mitzvah means son of the commandments, and Bat Mitzvah is daughter of the commandments. It is the same ceremony. Only one is masculine, and the other is feminine in the Hebrew language.

My students started out with me at the age of 12 and finished at age 13. They have now come of age in the Jewish religion. They were now able to be counted in a quorum of 10 called a Minyan, which is the minimum amount of people needed to conduct a service. This age is a turning point, especially for boys, who are turning into young men. They think they know everything, and they don't. Their behavior is often unruly. It's not an easy group. However, I had 40 students each year in my class, and the end result was remarkable. They could lead the Friday evening and Saturday morning service. They read from the Torah scroll, which is written in Hebrew calligraphy without vowels, only consonants, chant from the Prophets (Haftarah) and deliver the sermon. What an undertaking!

This afternoon, after I have heard each student read their particular portion that is read on the week of their Bnei Mitzvah (plural), we go to the sanctuary, and 1 or 2 get a chance each week to lead all or part of a service.

A student who will become Bar Mitzvah in about 2 months is our student of the day. He is very tall and lanky and towered above the other students. This student was not very popular. His size was not necessarily proportional to his emotional maturity. He is now chanting a prayer, which in Hebrew goes like this:

Yitgadal v'yitkadash, sh'mei raba

He begins by saying:

Yiskadu and Yiskadon't—Yiskawill and Yiskawon't

I have him stop and do it again. He says the same thing. In his favor, he may have gotten some help from an older relative who used the old Ashkenakic (Eastern European) pronunciation, which would have been:

Yiskadal v'yiskadash, sh'mei raba

HOWEVER, there is no do and don't—will and won't

At this point, my assistant Bar Mitzvah coach comes in and hears him. All he could do was shake his head. We straightened out the kid and finally got it right.

That is not the end!

The day of the Bar Mitzvah, it is an old European custom to throw candy at the Bar Mitzvah boy, when he has finally completed his task. It is a symbol of sweetness and is usually done with jelly-type candies that are individually wrapped. You can get these in bulk in the supermarket.

This morning was different. The family made up bags of sweets that contained several full size candy bars, gum, an apple and other small candies. They were tied up on the top and looked like a trick-or-treat bag.

After the Bar Mitzvah finished his duties on the pulpit, it was now time to toss the candy at him. Instead of opening the bags and tossing the small candies, people threw the whole bag at him. There were between 75 to 100 bags. People on the pulpit, including the Bar Mitzvah, were being pelted with missiles. The representative of the Sisterhood, who was going to make a presentation of a gift to him, was using the breastplate from the Torah as a shield to protect herself. I hid under the reading table where the Torah lay, and the Bar Mitzvah fell on the floor. It reminded me of the game at the fair, where you throw balls at fuzzy cats and try to knock them over. I have never seen such a display before or since.

When the mother was questioned and asked why she did this, her response was, "Oh! I didn't know that they were pieces of candy. I just heard candy, so I made up these lovely bags."

This now made it clear where he got:

Yiskadu and yiskadon't—Yiskawill and yiskawon't

People never cease to surprise me.

David Pinkwasser

I Thought the Roof Was Coming Down

The Persians are coming, the Persians are coming! No, it was not political demonstration from Iran, but rather a Persian wedding and premarital blessing with very special customs not known to most Ashkenazic Jews (Eastern European).

B'nei Mitzvot is not the only time we celebrate with a candy toss. It is also very common during an Ufruf. This is a special ceremony where the bride and groom come into the sanctuary on the Sabbath before their wedding. They are publically announcing their wedding to the congregation and receiving a blessing on the pulpit, from the rabbi.

It is also a custom to sing a happy song as they go back to their seats. It is usually:

Simon Tov and Mazel Tov

(Good signs and good luck)

People smile and clap as they descend and often through candy. Generally, the candy, as mentioned before, is wrapped jellies that are soft. Tonight was different. NOT LIKE THE BAR MITZVAH. THEY WERE NOT PELTED WITH ROCKS!

The bride was Iranian. It is a custom in Iran to not only throw soft candy but also to throw a Persian treat. I do not know the name of the candy, but it is small and looks like puffed rice with a white chocolate coating. Many of the women were shouting:

LO LO LO LO LEEEEESH

This is an expression of great joy. It was also repeated at the wedding when the bride came down the aisle.

I looked at my podium and saw the white candy on it. I thought it was the popcorn ceiling coming down because of all of the commotion and candy being thrown. I must have hit the ceiling, and now parts of it were coming down.

In a short while, I realized that it was the Iranian candy that was being thrown to wish the couple added sweetness. This was similar to throwing rice at a couple when they were departing from the ceremony.

It was quite a mess to clean up. The wrapped candies were picked up and eaten, but the custodian had a job ahead of himself cleaning up the white chocolate candy.

It is nice to see that we are all somewhat unique and different but the same in more ways than we know. As a rabbi and the president of the ministerial association, I have found that once we got past our different theologies, our goals, hopes and wishes were pretty much the same.

Interfaith Thanksgiving Service Lifelong Friendships

Lots of food: green bean casserole, candied sweet potatoes, turkey, stuffing, mashed potatoes etc. Family, friends and usually a prayer of thanksgiving. This was not a religious observance. Things have now changed.

This is something of which I was not familiar. An interfaith Thanksgiving Service. Where to begin? I got a visit from a Pastor, who we will call Tom. He arrives at the synagogue one day before Rosh Hashanah. This is like going to a church the day before Christmas and asking if they are interested in doing an event in the future.

I was expecting about 1,200 people at worship, and preparations were being finalized. I certainly was not interested in some service that had to do with a legal holiday. I have a hard enough time convincing people to come to services for a Jewish holiday, but this seemed out of the question.

I very politely told him that his timing couldn't be any worse, **but** he could show up next year a little earlier, then I'd be happy to talk to him. I never thought I would see him again.

It is now, one year later, about a month or so before Rosh Hashanah and here is Pastor Tom. He said,

"Hi. Remember me! I'm Pastor Tom from the So and So Church, and we are planning to have an Interfaith Thanksgiving Service. We'd really like you to attend."

I did say that I would speak with him. He explained that all the other congregations were Christian. If we did not join them, it couldn't be interfaith.

THE ADVENTURES OF THE RABBI WHO BECOMES A FLIGHT ATTENDANT

I listened carefully and consented to come to a meeting and hear more about it. I still had my doubts, but I was willing to listen.

At the meeting, I met a wonderful group of pastors who were very enthusiastic about the prospect of this service. I did warn them that I was not Christian, so there could not be any mention of Jesus in the service. Praying to God was fine, but being a non-Christian group, our people would be very turned off if Jesus was mentioned in the service.

We talked about our commonalities rather than our differences, as well as our view of the community to which we all served.

The service was to take place on Wednesday evening before Thanksgiving, and our wives were invited to process into the sanctuary with us and then be seated up front.

This all sounds great EXCEPT, we always went up to our vacation house in the mountains for the

Thanksgiving weekend. A lay person covered for me in my absence, and the kids were out of school. NOW, we could not leave early on Wednesday, but we would leave right after the service. My family was not happy about it, but I told them I felt obliged to do it.

Just when things couldn't get worse, I received an invitation to a reception after the service at Pastor Tom's house. Of course, I had to attend. We then decided that we would leave in the morning. No one was happy about it.

The service was beautiful. There were people from all different congregations together. Many of the attendees were neighbors, friends and others in the community that we had known but had never worshipped together.

On the pulpit, all of us were introduced, and the wives asked to stand up, and the atmosphere was beyond cordial.

Now, at the reception, we got to meet the others in a more casual environment. The kids seemed to get along great, and the food and libations were excellent.

While at the reception, the subject of skiing came up. My wife told the pastors that we had a home in the mountains and that maybe after the holidays, those who wanted could get together and ski.

Shortly after New Year's, I arranged the trip. The guys said that they thought they would never hear from me. They did. The trip was a great success.

When we returned, the 3 pastors who joined me were very sore. I told them that they were out of shape. I go to the gym at least 5 times a week. One of them said that he was interested in doing that. Could he come as a guest? I responded, "Of course."

Tom enjoyed the workout and became a member. Shortly after, the other 2 joined. Now, for the next 5 years or so, we met at 7:30 in the morning and by 9:00, we were back in our respective offices.

While at the gym, there was a reporter from the local newspaper. He saw a story here. We now had our pictures plastered all over the community with the line:

A HEALTHY SPIRIT AND A HEALTHY BODY

Ever since that event on Thanksgiving, we have become lifelong friends. We have shared the happy, sad and social aspects of each other's life. We can count on each other, no matter what the circumstances might be. This wouldn't have happened without the Interfaith Thanksgiving Service.

BUT WAIT THERE'S MORE!

At the service, there was an offertory. This is something very strange to Jews. On the Sabbath and Holy Days, money is never exchanged. People become members of a congregation and pay dues. It is an annual fee that is usually paid monthly through billing. No one is ever turned away. If they cannot afford the dues, the amount can be altered to meet their particular budget.

My kids thought this was great. They've never seen a basket of money float around. I gave them a few dollars to put in. My youngest wanted to know if he could take some out since there was so much in there. I told him that he ABSOLUTELY could not!

When the basket came back, there was $400 in it. The money was given to a social service agency that distributed it to those in need on our behalf. This amount of money lasted until the end of January. Then, we had to go to each congregation and beg for more. I told the group that I was more familiar with billing, but on Yom Kippur, the Day of Atonement, we do an appeal. We have pledge cards with everyone's name on it. There are tabs to bend over and then place in an envelope. They have until the end of the year to make good on their pledge.

I told them that next year, I would like to do the offertory. I stood up there and said proudly:

Ladies and gentlemen. In a few moments, we will begin the offertory portion of our service. The money that you place in there is not going to my Hawaii fund (Chuckle, Chuckle). It is not going to any of our congregations. It is going to help poor families that may not be eating a turkey tomorrow or they may be eating at a soup kitchen if you were going to put a dollar in the basket, put in two. If you were going to put in two, throw in a five-dollar bill. When you sit down to your Thanksgiving dinner, you will know that you helped a poor individual who may not have had anything to eat.

The basket went around, and $1,400 came back. The group quietly turned to me and said:

You're President!!!

I was now president for the next 13 years.

As president of the ministerial association, I looked at the cash distribution. It was as follows:

$10 per car load for gas (This is when you could fill your tank with $10)

$10 each for groceries (Given food vouchers for a local supermarket)

$4 each for fast food (Voucher for local fast food restaurant)

The money went further, but it wasn't a great plan. This is what I came up with.

We have 24 congregations. 12 will bring 30 bags of groceries from your congregation. Pick a month, and that will be your assignment. Also, $100 in cash. If you have congregants that don't have time to buy groceries, they can put a few dollars in and getting to $100 isn't hard.

The other 12 congregations will do homeless packages. This is a ziplock with soap, a wash cloth, shampoo etc., so that when the homeless are at the Armory in the cold months, they can have toiletries to clean themselves. Also, $100 in cash.

We then split the group into north and south services. The north service brought in $1,100, and the south still brought in $1,400, even without the other 12. With a little politicking, I was able to bring our budget up to $10,000, and the output was much less because now we had a food bank and didn't have to rely on supermarkets or fast food restaurants.

To say that I was proud of what this group was able to do would be an understatement. Our community has a lot of which we should be proud.

THE FUNNY SIDE!

There is a lot more than just the serious stuff! Now that I knew the group well, I would purposely do some funny things. If the group decided to wear black robes, I'd wear white. If we decided on white, I'd show up in black. (I'm really trying to get over my shyness—Huh!) The Methodist Pastor would say something like,

Once again, his "Rabbiship" did not get the memo!

Everyone laughed. The wives were tricky, too. When they were introduced,

"This is Rabbi Pinkwasser and his lovely wife Ann", then someone else's wife would stand up.

Half the room was hysterical, and the other half was confused.

"This is Pastor Tom and his wife, Jane." At this point, Ann would stand up. It added some levity to the evening.

Ann also wanted to do something for the wives. She wrote a newsletter called the Forever United Clergy Kin. The title is a little obscene but comical. If you look at the title, you'll see some nasty language. It was a spoof, a parody. Articles about the group were written that were funny and exaggerated. She did several a year. The wives couldn't wait for the next issue to come out because they were so funny. She is a much better writer than I am! She kept them in stitches.

You never know where life is going to take you. If you don't jump into things and try it, you'll never know. Life is not lived in front of the TV set and lying on the couch. You need to be out there, try everything and see where it takes you.

THE ADVENTURES OF THE RABBI
WHO BECOMES A FLIGHT ATTENDANT

I Want A Passover Seder, Too!

As a result of the Interfaith Thanksgiving Service, there was much interest in Judaism, especially around Passover time. Since Easter and Passover are often very close in time, people now felt a little more comfortable about approaching a rabbi and asking if he could do a Passover seder for them since the Last Supper was supposed to be a Seder.

A seder is a ritual meal. It has symbolic food that is eaten to recall and relive the exodus from Egypt (Much more accurate than Charlton Heston's portrayal). It also has a big dinner of traditional Jewish food, which is eaten at this time. All of the food is to be free of leaven. The Passover Story tells that the Jews left Egypt in such haste that bread didn't have a chance to rise. Therefore, during Passover, any leavening agent or grain that can self-leaven is forbidden during the 8 days of the holiday. The food is quite unique.

I have now been asked by several churches to do a seder at their facility and explain the significance. This is known as a model seder. I would have all of the ritual foods but not provide 4-course meals for a group of 50.

This went very well for years. I was happy to explain my traditions until it got out of hand. Now, so many churches were asking for this ceremony that I couldn't handle it. I came up with a solution.

I would do one seder at the synagogue and have 100+ people in attendance. This worked well. THEN, people said, you talk so much about the traditional food, as well as the ritual food, we would like to experience it. I then asked the Sisterhood of the congregation if they were willing to undertake this task. We charged a small fee for the meals and fed about 100 people. Pretty cool!

Now, even with this one big seder, it didn't end here. As a rabbi, I also visited the prisons. I did several seders there. The preschool at the Jewish Community Center had me do one every year. The Hebrew School at the synagogue had me do one. Of course, at home, I had two more to do. By the end of Passover, I was severed out!

It was truly a gift to be out to share my traditions and customs with others. I was also pleased to be able to learn more about other's customs. Knowledge is a beautiful thing. There can never be too much. It gives you perspective and understanding of the world around you, what people believe, and how they act. You and they are enriched. It adds balance and comfort to your life. You know where you are in this world and know what direction you are heading. I am truly blessed to have these experiences and hope that those that I have shared them with also feel blessed about expanding their horizons.

David Pinkwasser

The Rabbi Leads Services At the Methodist Church

(Are You Confused?)

I have received a request from my colleague, Pastor Jim. He tells me that there is a conference that he would like to attend of all Protestant Ministers. Most people in the community are attending. Here's the problem. There is no one to cover for him because they will all be there. Would I? I ask,

"Jim. Are you confused? Do you know who you called?"

The response is, "Yes."

"You can pick the sermon topic, and there are hymns and prayers that are non-Christian that will not go against your beliefs."

I consent and then go to the service.

The opening hymn is Shalom Chaverim. It is a Hebrew song of which I am very familiar. I mean, Hello, all my Friends. A good way to start. HOWEVER, the Methodists did not know that the Ch is guttural, like the word **jota** in Spanish, not Ch, like a **chair** in English. I had to laugh to myself as I heard the people sing. It was so thoughtful of them to make the effort to make me feel comfortable, even though they were mispronouncing the word.

The concluding hymn was called:

The God of Abraham's Praise

I never heard this hymn before until the organ began to play. It was **Yigdal**.

The words were different, but the tune was familiar. It is a concluding hymn sung on the Sabbath. As I looked at the hymnal, the footnotes said that this hymn was designed after the Jewish hymn Yigdal.

David Pinkwasser

The morning was so meaningful to all of us. Each of us reached out further than we needed to make each of us feel comfortable. It is a lesson in sharing traditions and lives for which I am most grateful.

I Brought Jewish Special Education to Arizona

I moved to Arizona in 1975. When I was ordained, I promised my wife that I would not have a congregation but would serve as a Jewish Chaplain in a Care Center for the aged. HOWEVER, the war in Vietnam changed those goals. Non medical and essential services were not being served. Congregational life seemed very political to me, and working in a facility seemed more regular in schedule to me and the family.

In the meantime, I began teaching high school in New York. I taught Spanish, Italian, ESL, and later in Arizona, Hebrew. When I got to Phoenix, there were no openings in foreign languages, but there were lots of openings in Special Education. I had no idea what that was because, in New York, it was not available in most schools. I asked my friend what it was. I was told that you would work with kids who had learning difficulties. I said:

"That sounds good. I can do that. It would be very challenging."

I immediately signed up for university courses in summer school and got a provisional certificate and with another year of night classes, became certified in all areas of Special Education.

I had a cross-categorical resource room and worked with kids who had Learning Disabilities, Emotional Handicaps or Mental Handicaps. It truly was rewarding and very successful.

Two months later, I was taking my nephew for Bar Mitzvah lessons at a large synagogue in Phoenix. While I was waiting for him, the religious school director said that he and the synagogue wanted to start a Special Education program for kids who were dropouts and potential drop outs of Hebrew school. They were looking for someone with both the Special Ed. Background and Judaic-Hebrew knowledge. He looked at me and said,

"You're it! We want you badly. Name your prices!"

Needless to say, I was shocked. I thought about it and found this to be a new avenue to explore.

I was done teaching by 2:00. This program would be from 4:00-6:00, twice a week. I loved it. I got to use more of my skill sets and help kids that needed me.

The program was so successful that it was expanded to four days a week from 3:00-6:00. It was so successful that people were leaving other synagogues to join this one because they had this program.

One of the people from another synagogue asked if I knew anyone who might be able to find someone who might be able to do what I do for them. I had a colleague who was Jewish and a Special Ed.

Teacher and thought the program was exciting. I trained him, and now 2 synagogues had this program.

In time, the programs expanded and expanded. Almost every synagogue in the Phoenix area has a Jewish Special Ed. Program. There is now an organization with a full-time director that not only supervises the educational programs, but also provides group homes and jobs for people with Special Needs.

I am so happy to see that the little class that I taught has blossomed into not only a tree but an entire garden of services for this segment of the Jewish community.

It is now 47 years later, and those who attend Hebrew and Religious Schools see the Special Needs class as a normal part of the school curriculum. I am proud to say that I started it.

THE ADVENTURES OF THE RABBI
WHO BECOMES A FLIGHT ATTENDANT

The Rabbi Faints on the Pulpit

How could this happen? There was no blood or gore. The rabbi wasn't sick. It has never happened before, and now here we are.

I must first tell you that **it wasn't me**. It was my colleague. Let me set the scenario for you.

It is the end of the Yom Kippur service. This is the most solemn day of the Jewish year when people ask for forgiveness of their sins. The services are long and tense and last for hours the night before and then the next day from morning until sundown. There is no eating or drinking. (Remember this. It will come in handy shortly.)

At the end of the service, as a sign of complete submission to G-d on behalf of the congregation, during the Aleinu (One of our concluding prayers), the rabbi prostrates himself in front of the ark containing the Torah scrolls.

I am now 28 years old. My colleague, the Senior Rabbi, is 68. This doesn't sound like much, but if you go back 45 years, a 68-year-old is not the same as one now. He looked like an old man. He was a grandfather figure, where as I was a father, or possibly an older brother figure. The rabbi, whom we will call Jacob, was older than my parents.

After the prayer is concluded, I get up, and Jacob is still on the floor. I whisper:

"Jacob, get up!" There is no response. I say it again. We turn him onto his back, and his eyes have rolled back into his head. He was just muttering sounds. We don't know if he is having a stroke or a heart attack. We ask if there is a doctor in the house. This is pretty funny in a Jewish congregation. About 10 people come running up. We call 911, and the paramedics come and take him away on a stretcher.

I finish the service and we are all concerned about the condition of the rabbi. It turns out that this is a result of dehydration.

You must understand that traditionally nothing is taken by mouth on Yom Kippur, even water. This is what I learned at Rabbinical School in New York. HOWEVER, we are in a desert in Arizona. The rules are different. No fasting is to cause physical damage to our health. If you are on medication or have a physical condition, you may eat and take your medicine, and this is not only allowed but encouraged. In this case, there was none of the above. HOWEVER, we have learned throughout the years, with a large Jewish population living in the desert, that water is important to keep us from getting ill. Present case, to be exact. It is not a time to be drinking martinis or other fancy libations, but some water, in small amounts is encouraged. This was a lesson to be learned.

I have found that not only the water but some drinks that have electrolytes will help keep you from passing out.

It is amazing the knowledge that we learn as we go along. Sometimes, the lessons are painful and require medical assistance, but we learn a lot from our experiences. Life is a journey. We constantly pick up information on this trek and need to file it into our brains for future use.

*THE ADVENTURES OF THE RABBI
WHO BECOMES A FLIGHT ATTENDANT*

Don't Shoot—I Washed My Hands!

I was standing at gunpoint with my hands in the air. I was not taken hostage. The men with the guns were good guys. What's up?????

One of my duties as a rabbi was to visit the Jewish population in prison. It was a contract that I had for 13 years. It was very part-time. The contract required an 1 ½ hour session at each facility. If prisoners could not be put together into one unit, I often had to do multiple sessions in each section. I did this 5 days out of the month. I drove all around the state and tried to make a circuit so that in 2 days, I could cover most of the facilities.

The question that people ask:

"Jews in prison. Oh, come on! They must be bad check writers."

The truth is that half of the people were white-collar, but the **other half—were** tough, dangerous criminals.

The next question that people ask is there must be only a couple of people. Jews made up ½ of 1 percent of the population. In the general population, we make up about 4-5%. Nice to see that we are underrepresented.

At the prisons, I would do a service, Torah study, counseling or Judaic teaching. It varied as to the makeup of the population. Some were not interested in studying, and others were not interested in a service. I had to play to my audience.

Today, I have done a service and Torah study and am ready to drive home. It is about a 3-hour drive. Before I leave, I make a trip to the restroom. It is one seater, located in the chapel and is only used by staff. Here's the rub! This prison, which was the only one, required me to hear a body alarm. I looked like a receiver to a telephone (when we still had dial phones) that had a button on it that was a silent alarm. I always kept it in my back pocket. Now I'm in the bathroom. I drop my pants, and the alarm hits the floor and goes off. I have no idea that this is happening. I casually finish, wash my hands and prepare to leave the facility. As I open up the door, there is a guard with a rifle pointed at me. I raise my hands in a "Don't Shoot Position," and I say to them:

"DON'T SHOOT. I FLUSHED, I WASHED MY HANDS. WHAT'S GOING ON!"

They then tell me that my body alarm went off, and now the ENTIRE prison is in lock down. We thought you were being held captive. I didn't know what to say. Everyone was glad that I was safe, but the catastrophe that I caused affected everyone in the prison. I was warned by the warden that if this ever happened again, I would be gone. I assured him that it would and learned a great lesson.

Prisons can become dangerous places. I had been there so many times that it was another of the many venues that I served. I was very nonchalant about being there. After this experience, I always looked around with more caution and definitely protected that body alarm.

David Pinkwasser

Teens Polish the Cemetery

(Please Let Me Do It—It's My Turn)

I made a little white lie. I told you that there would be no more funeral or death stories. This one is worth telling and you'll see that it is not morbid.

As the rabbi of the congregation, I would talk to the youth group, all high school age, about subjects that they needed to know about, not just Bible Stories. I talked to them about a recent death in the congregation, which the kids were very curious to know, and then took them to a mortuary.

The funeral director, who was Jewish, talked about the ritual that we do to prepare the bodies for burial, which does not include embalming. It is a washing of the bodies to symbolically wash away sins since the deceased are no longer able to do that on their own. The body is dressed in a shroud, which is a simple burial garment so that on the Day of Judgement, you are standing before G-d on your good deeds and not impressing anyone with designer clothes.

After the lecture, the kids were taken into the prep room. They were all very skeptical as to what was going to happen next since the only experience they had was from scary movies. The funeral director whispers to me that there are no bodies there at this time, but don't say anything to the kids.

Everyone gingerly walks in and finds a room that is brightly lit with light colors on the walls, with the appearance of an operating room rather than a scary dungeon. The funeral director informs them that this is his workplace. I want to be in a pleasant environment with lots of light to see what I am doing. When we leave, the mystery is gone, and the teens feel very comfortable being there. I wish that I could have done this trip with their parents!

Let's Continue To The Cemetery

Now that the teens have had this initial orientation, I asked if anyone was interested in coming to the Temple Cemetery before the High Holy Days to polish the bronze markers on the graves.

Although the mortuary visit was successful, the number of volunteers was miniscule. I encouraged 4 of the youth to do it, and they reluctantly consented.

At the cemetery, they saw the names of people that were familiar to them. They asked me who that was. They have the same last name as one of my friends. I walked with them around the cemetery, telling them stories and the history of many of these people. I knew all of them and their families. The kids found it fascinating. We polished up the markers with orange oil, which brings back the nice bronze appearance that is stained by the hard Arizona water. There was a sense of accomplishment when they saw the beauty of their job and learned a lot about the history of their congregation, community and their friends.

When the teens came back to the temple, they talked all about their experiences. The others were jealous and sorry that they did not go. The next year, people were begging to be part of this group. There were over 20 people that wanted to do it. We had to limit the number. Now we had a first come, first call activity that everyone thought was so cool.

It would be great if you could make a cemetery a cool place. They were able to look beyond the obvious and were able to share this experience with others. This is maturing Jewishly. They were now ready to be one step closer to adulthood.

David Pinkwasser

The Tefillin Police

I am giving you a disclaimer now: there is no such thing as Tefillin Police!

Before I begin this adventure, I need to explain what tefillin are. They are translated into English as phylacteries. Now, is that clear? I have no idea, nor does anyone else, what a phylactery is. Sounds like a medication or an illness. Tefillin are small wooden boxes and contain a prayer that speaks of the Oneness of God. They are attached to the arm and head with leather straps. The commandment to bind ourselves in the tefillin comes from the book of Deuteronomy in the Torah. Most traditional synagogues encourage this practice on a daily basis as a means of symbolically binding ourselves to God, which is done daily. It is never done on the Sabbath, called Shabbat, because this is a holy day already, and the need to bind ourselves to a holy object is not necessary.

THE ADVENTURES OF THE RABBI
WHO BECOMES A FLIGHT ATTENDANT

In Reform congregations, this practice is seen less and less. HOWEVER, our story now begins in summer camp with our teen group. There is a week up in the mountains when all the reform congregations take the teens to the mountains for a week-long retreat. While in the camp, I woke up in the morning and found a quiet place in the corner of an empty cabin to do my morning ritual. We came to notice that all of the rabbis, except for one who did not come from a traditional background, were doing this. We decided that we would share this experience with the youth. We invited anyone who wished to join us at 7 am the next day, on the back porch, for a service with tefillin and an explanation of how it is done. No one was required to attend. This was strictly voluntary. The next day, about 35-40 teens showed up and were very interested. We did our explanations and invited them to join us. The kids loved it. This was a totally new experience that they had read about in the Shma prayer, which comes directly from Deuteronomy but had never experienced.

Months go by, and I get a call from the URJ. This is the Union of Reform Judaism. They asked if this was Rabbi Pinkwasser. I told them it was. Then they said:

"We understand that you lay tefillin every day."

I said to myself, what is this? Is this the tefillin police? Do they think I am too traditional to be part of a liberal reform movement? My response was:

"Yes! What's it to you?"

They then explained that in the upcoming convention, they would like me to lead a Sunday morning service with tefillin. They also wanted me to explain the significance, as well as encourage people to bring them if they had them.

You must know that a Sunday morning service usually gets about 10 people to show up. The big service is on Shabbat with several thousand in attendance.

I consented to do this, and to my surprise, about 75-100 people showed up. It was a very moving experience, and many felt enlightened by this new knowledge that they had heard about but never actually took part in.

After the service, many people came up to me and were grateful to have another pathway to binding with God and spirituality. This practice, which for many years was dropped from reform practice, is now coming alive with a new interpretation of its meaning.

For over a year, I periodically received correspondence from people who were now engaging in this practice and found it to be very fulfilling.

I was so blessed to have these experiences to share them with others, and to know that there are no TEFILLIN POLICE!

David Pinkwasser

When Singing Is Inappropriate

Music is good for the soul. It is soothing or exhilarating. There are times when you could be wrong, dead wrong!

These are 2 vignettes about singing at the wrong time. You might have an idea in your mind when that might be, but I think I can top that.

The first time is at a Bat Mitzvah reception. It was very beautiful, at a nice hotel and everyone was very dressed up. The grandfather of the Bat Mitzvah was from Eastern Europe and had a European accent. He was a retired cantor from the old school. What this means, they were more of an operatic singer than a song leader or crooner.

Without warning, he goes up to the band and hands them sheet music. They have never seen this music before. He said I want to sing a song to my granddaughter. In his European accent, he began:

Sonya, I love my Sonya

And you studied Toyra (European pronunciation of Torah)

You studied Toyra!

You studied Toyra!!!

It kept getting louder and more intense.

By the third Toyra, I couldn't control myself from laughter. HOWEVER, it was not nice for the rabbi to laugh at this old man, so I went into survival mode. I began to cough and then excused myself out to the corridor. When I got there, I saw about 10 people also fleeing. There were old-time congregants, board members and prominent people in the community. We all could not sit quietly, watching this spectacle, which to everyone there was inappropriate, but to him, it was perfectly normal.

As a rabbi, you are a leader. Proper behavior is always expected. I think I did the right thing.

The second one was at a funeral. It was my uncle. The family asked if they could play a song that he liked at the funeral. This is highly irregular, but it's family. You make exceptions.

My cousin walks into the cemetery with a boom box and says that the song is a Bobby Darin song. I must say, I like Bobby Darin, but at a funeral, not so much. I told them that as people were leaving the cemetery, we would play the song.

Now, the funeral has ended. The song is played and I dismissed the crowd. My aunt now stands up and starts swaying with the music. She then starts singing. THEN, she turns around and tells everyone to sing with her. I was dumbfounded. I have never seen a spectacle like this before or since. Then my aunt said to the funeral director:

"I bet you never saw anything like this before."

He responded, "No, I haven't," shaking his head.

She had no idea that this was not a compliment.

Music can sometimes be inappropriate. Spiritually uplifting music and hymns work well, pop music, not so much.

You never know what people will do!

David Pinkwasser

The Mile High Club
Is It Open to the Public?

How can I get a membership? Is it real? Do people really do this on planes?

These are the questions that a flight attendant would be asked once you knew them well and each has had a number of drinks. For those of you who may think that this is a frequent flier program to get you into first-class—It's not! The mile-high club is the performance of sex at 40,000 ft. It is secretive because it is not allowed on any commercial airline and has the mystique of being not only exotic but naughty. You know that you are the only ones on this flight doing something wrong and getting away with it.

I will tell you that in my 22 years of flying, I have never been a member of this club, even though I have met several members. This tale starts with a Jewish overtone.

THE ADVENTURES OF THE RABBI WHO BECOMES A FLIGHT ATTENDANT

I was working as A Flight Attendant. I greeted people as they entered the plane and cheerfully welcomed them on. A couple came on who were very distinctive. The woman wore a pink leather jacket with a kind of boa, also in pink, attached to the collar and down the front. She looked like a walking bottle of Pepto Bismol. She was nice-looking, not a glamour girl, but also not a great critic of fashion. She was accompanied by a man who wore a t-shirt that had a logo of a Jewish dating service. I, being Jewish and a rabbi, made a remark about it, but he seemed not to pick up on any of my innuendos. Okay, not the first time that this has happened.

When I took their drink order, the woman wanted a frozen daiquiri. I reminded her that she was on an airplane and not at a resort. I didn't have a blender. She could have a vodka cran with a twist. She was fine with that. He ordered 2 beers. This was before credit card payments on flights, so she took out her wallet and paid for it all. I thought that this was strange. A Jewish guy trying to impress his date would never let the woman pay for drinks. What do I know? Old school!!!!

When I went to serve them, they were gone. I looked in the aisle and around the plane. We were ¾ full that day. They were nowhere in sight. I went to the back of the plane and asked the other 2 flight attendants if they had seen the "Pepto Bismol Girl". They said, "No". The only place that they could be was in the back lav. I banged on the door, and the 3 of us stood there with our armed folded in front of us. The woman was sitting on the lid of the toilet and was licking her lips. The man was zipping himself up. They smiled with a devilish grin and proceeded back to their seats. The B Flight Attendant told me to make my announcement:

LADIES AND GENTLEMEN: WE'D LIKE TO WELCOME THE NEWEST MEMBERS OF THE MILE HIGH CLUB. LETS GIVE THEM A BIG HAND.

The passengers went wild. There was clapping and cheering. The guy had his hands clasped over his head in a sign of victory. The woman was very quiet and looked a little embarrassed. I then served them their drinks, and we continued.

Upon landing, I welcomed everyone to Cleveland and reminded them of our frequent flier club that would get them free fights. The man then got up and said, "Don't forget about the Mile High Club?" I then repeated what he said, and the crowd went wild.

As the couple left the plane, the man walked up to me and thanked me, then handed me his card. I quickly looked at it. He was from a Stud Service. The woman had hired him to do the mile-high thing to probably live out a fantasy.

When we got to the hotel, we got on the internet to check it out. He and 5 other guys had a service. It was not even called an escort. It was raw sex. They had photos and descriptions of what you could purchase. This was very different than the female ones where provocative clothing was worn and innuendos were made, but not blatantly displayed. This service left nothing to the imagination. You live and learn.

PS

Several weeks later, I was in my position, and an FBI agent came on the plane. It is my job as the Flight Attendant to check the IDs of Law Enforcement Officers (LEOs) who were carrying weapons on the plane. I asked the FBI agent why this service was allowed to function without being undercover or, at least, discreet. It was obviously services that they were selling for money that violated the law. Their response was this:

We are interested in catching hijackers and terrorists. These guys that were selling "A little chicken on the side", were of no interest to us. Rather shocking, but as I've said before, "You live and learn".

*THE ADVENTURES OF THE RABBI
WHO BECOMES A FLIGHT ATTENDANT*

Nice Flashlight, Buddy

"Nice Flashlight, Buddy!" This remark comes from one of my fellow crew members. My response is, "Thanks?" I look down on the counter and my flashlight is missing the bulb and light part. I only have the base. Doesn't sound like a big deal, WELL, let me tell you!!!

I am a new flight attendant. I've been on the job for several months and am still on probation. That lasts for 6 months. One of the FAA and company requirements is that everyone has a flashlight in case of emergency, etc. There are flashlights in the front and back of the plane, but if something drastic happens and you are not in that spot, it behooves you to have it on hand. (This is years before they were part of our cell phones.)

We also had briefing days, which would pop up unexpectedly and unannounced, where you have to go through an inspection of the uniform, manual (totally updated) and flashlight. You even needed to bring it to recurrent training as part of the inspection. This was serious stuff.

I somehow lost the top part of my flashlight and now had no way of getting a replacement. I was already on the plane. I worried that if a supervisor came on, unannounced or an FAA inspector, I would be fired.

I have to fly 5 legs that day. Numbers 1-4 go smoothly. On leg 5, an FAA inspector came on the plane. This was a rare occurrence, but, my luck, here he was. At the time I did not know that he was actually commuting from one city to the next and really didn't care about my flashlight. In reality, an inspector, even if he is commuting, should go through the inspection, but he didn't.

I now offer him free drinks, hoping that he will get tired and fall asleep. Of course, he is not allowed to drink while on duty or commuting, but I didn't know this at the time. I was still pretty green.

We get to our destination for the evening. It is Midland, Texas. I have heard that it is a sleepy little town. When I get there, I see many high-rise buildings and a large downtown area. We stay at a large hotel which has a lot of amenities. HOWEVER, there was a lot that I was going to learn.

When oil was king, Midland was a bustling city. When prices dropped, it became practically a ghost town. I found out later that only 2 or 3 floors of the hotel were being used because there weren't enough guests to fill it. I don't see any guests in the lobby and now head for the gift shop to look for a flashlight.

When I spoke to the clerk in the gift shop, which had no customers, she informed me that they didn't have flashlights but did have Brighton. I asked what Brighton was and found out that it was a nice brand of costume jewelry. This was not to help me. I asked where I could find a flashlight, and she told me there was a chain drugstore about a mile away, so if I liked walking, it wasn't too far.

I thanked her and began my walk. The street was deserted. I was the only one walking by closed stores and no people. It was very spooky. I then passed by a man, who was walking with an attaché case and stopped to greet me and ask how I was. He probably was so happy to see a live person that even a stranger was someone to greet and talk to.

I continued my walk, which took me almost an hour. I found the drugstore and got the flashlight. It took me another hour to get back to the hotel. I learned that this was a "Texas Mile."

That evening, I met the crew and we went to dinner in a small bar that was a block from the hotel. There were no cars in the street and no cars in the parking lot of the bar. We opened the door and found it filled with people and a lot of chatter and liveliness going on. I thought I was on Candid Camera! Where did all these people come from and why not a lot of cars? It was so strange. I did find out that this was the premier place to eat in Midland. Everyone congregated there, and the food was good.

I must say that anytime I had a Midland overnight, that was the place I went to. I even spent one Christmas Eve there, and it was just as lively and happy as it had been before.

P.S.

I had a friend that I liked to fly with. She was always afraid of going to Midland because she heard these creepy stories. She saw that I had a trip there, and there was an open position on the crew. I informed her that it would be a good trip and that she shouldn't be frightened. She told me that the last trip that she had there was when she was on reserve, so she had to take the assignment. She went to her room and put chairs in front of the door as an extra added protection. I assured her that this was not necessary. We both liked to work out and enjoy good wine and food. She asked it I wanted her to bring a bottle of red or white. I opted for the red. I told her I would bring appetizers.

The gym at the hotel was on the second floor next to the pool. Needless to say, there was no one else at the gym or anyone at the pool. After the workout, we enjoyed drinks poolside at our "private resort." She thanked me for taking away her fears and found it to be a cool 3-day trip.

Things are not always as they appear on the surface. Sometimes, you need to delve further and realize you can enjoy something that others shun.

*THE ADVENTURES OF THE RABBI
WHO BECOMES A FLIGHT ATTENDANT*

Is That A Flashbulb Or Lightning?

BOOM—FLASH—BANG—SCREAMS

This is what I hear and see. I have no idea what it is. I've been on the job for 3 weeks, and now something horrific happens. Here's how it went.

I, being new, shared the jump seat in the front for takeoff and landing.

I prepared drinks in the back, sharing the galley with another flight attendant. This is pretty standard procedure for new flight attendants. You are always around someone more experienced to help guide you.

We are now about ready to land in Phoenix. The flight attendant in the rear would like to jump off in Phoenix and get a pizza. She asked if I wouldn't mind landing in the back, and she'd take my seat in the front. I said, "Sure." This was very exciting. It's the first time that I have landed alone. I think it's pretty cool. There are a lot of clouds outside, and it's a little bumpy. No sweat. All of a sudden, I hear a bang and a boom. I see a flash in the cabin. The 2 upfront could not see it because there was a bulkhead separating us. I thought someone might have used a flash to take a picture. Kind of odd since it was daytime. This was also before everyone had cameras on their phones. I thought there might have been an explosion. I did all of my flight attendant safety checks. I examined the back galley. None of the circuit breakers were popped. I felt the lav door with the back of my hands, as we were taught in training to check for heat and possible fire. All was good. I slowly opened the door, and all was clear. I now thought that I should call the cockpit. What was I going to say? I've never communicated with the cockpit before. The other flight attendants did that and then told me what they said. I now pick up the phone, and then an announcement comes on the PA. THIS IS YOUR CAPTAIN SPEAKING. YOU MAY HAVE WONDERED WHAT THAT NOISE

WAS. IT'S ONLY A LITTLE STATIC DISCHARGE FROM THE CLOUDS. NO PROBLEM. WE'RE HEADING DOWN TO PHOENIX.

When we landed, the provisioner came on the plane to stock it with drinks and snacks. He told me that we were hit by lightning. Come out on the truck. He then pointed to the tail of the plane. That burn mark is from lightning.

The first officer took me on the tarmac. Look at the engine. All of the rivets were knocked out of the housing and there was a burn mark down the side of the engine. I then told him what I saw and then what the provisioner showed me. We deduced that the lightning hit the engine, went down the aisle and exited through the back ceiling, causing the burn in the tail.

If I had been standing in the aisle when this happened, I might have been killed or injured. The plane was taken out of service for inspection.

We learned later that 2 more planes were also struck at the same time while trying to land in Phoenix. All were taken out of service.

Pretty exciting! And to think that this was my 3rd week. I probably should have quit at this time, but then I wouldn't have had all those other adventures.

When incidents like this occur, it is important to listen to the instructions of the pilots and flight attendants. We are there for your safety, even though you think of us as servers. If people are seated and strapped in during turbulence and other incidents like this, then you are safe.

I will tell you that this has never happened to me again, but if it did, I'd be prepared.

*THE ADVENTURES OF THE RABBI
WHO BECOMES A FLIGHT ATTENDANT*

I Think I Smell Smoke! Is It A Cigarette Or A Fire?

Sniff! Sniff! Something is burning. On an airplane in flight, this is not the odor you wish to find. After careful investigation, we determine it is coming from the back lav. After deeper scrutiny, we determine that it is someone smoking in the bathroom. This is strictly forbidden on domestic flights and is punishable with fines and imprisonment.

Knock! Knock! on the door. "Hey, what's going on in there." The toilet is constantly flushing. It is an air-powered toilet that makes a very distinctive sound. No response. We know that if it is a bowel problem or gastric problem, the toilet might be flushed several times, but not 10 or 20 times in a row. We know what is going on. The person in the lav is smoking and blowing the smoke down the toilet so it doesn't set off the smoke alarm. We now announce our entry. The door opens, and the room is filled with the smell of burnt cigarettes, but no cigarette to be found. It was obvious that the person threw the cigarette down the toilet. It's not a great idea with chemicals and sewer gas. Hopefully, it was extinguished. Who knows! It could have also been thrown down the trash bin, which has paper towels in it and could have caused a fire. The person was silent and wouldn't tell us, even though it concerned the safety of the other passengers and herself in the process.

The passenger was one whom we had spoken to earlier. Her mother had passed away. She and the family were heading out to make funeral arrangements. There was a lot of stress, and she— JUST NEEDED A CIGARETTE TO CALM DOWN! We informed her that even with stress, loss of a family member etc., smoking is not allowed on domestic commercial flights. She knew this. This is an example of someone trying to "Push the envelope" as far as she could. After much deliberation, we let her know that, due to the circumstances, we would not inform the police. If she tried it again, we would.

Sometimes, you need to know how to handle sensitive situations. Being a clergy helped us deal with this one.

David Pinkwasser

We're Gonna Doy! We're Gonna Doy! (Die in Brooklyn)

WE'RE GONNA DOY! WE'RE GONNA DOY! These are the words I hear shouted as we hit clear air turbulence. This is turbulent that is unexpected, and can happen at any time. However, planes don't fall out of the sky because of this. You get jostled around and even thrown around. It was pretty bad. I was in the rear part of the cabin, about 7 rows from the back. There was no time to get to a jump seat. All of the passenger seats were occupied, so I sat on the floor and held on the armrest. This is standard procedure. If you are on the floor, you won't fall and hurt yourself and possibly others. While on the floor, I hear again,

"We're gonna doy! We're gonna doy! Moishie lost his yarmulke (Jewish skull cap)". It was a group of Orthodox Jews who were flying from New York to Fort Lauderdale. My luck! It had to be one of my people who were hysterical!

I informed the group from the floor that besides being a flight attendant, I was also a rabbi. I want to assure you that all is okay. If you are belted in, you'll be safe. As for the yarmulke, it is somewhere here. This is an enclosed space. It didn't fly out the window. I explained turbulence to them and gave them a few reassuring words, and all was okay.

We landed. No one was hurt, and yes, Moishie found his yarmulke.

Another story, totally unrelated, but one that had my rabbinic background as a helper.

A woman from India was traveling alone. She was afraid of flying. The woman next to her was very caring and kept her calm and explained all of the little bumps and jolts that happen during a flight. At the next stop, the new friend left, and I was now in charge of a woman from India.

While still on the ground, I was having a conversation with my fellow flight attendant. The fact that I was a rabbi came out. The Indian woman piped in and smiled. She asked if I was really a rabbi. I answered in the affirmative. She responded, " Oh, Good! Now I feel safe!!" I replied, "Aren't you Hindu?" She responded, "Yes. But the Jews are God's chosen people, and you are a leader of them. Therefore, I feel very safe. All the fear of flying left, and she was confident that she would arrive at her destination unharmed.

You never know where life will take you and never know what people are thinking. This is the last thing I would have ever pulled out of my sleeve to calm a passenger down, especially one that I did not think was Jewish. The other group that was a whole different story.

You have to know how to play to your audience. Those who are business travelers are very different than those going to Cancun or Vegas.

Bachelor/Bachelorette Parties and people in mourning are totally different. You need to feel out of your crowd and play to your audience. HOWEVER, sometimes your audience is unknown, and you need to be ready for that. This was certainly the case, so you need to be ready for anything that comes your way.

David Pinkwasser

I Left My Heart in Havana

I know that I will never die because I do not have a heart. My heart is interred in Cuba, and that is where it will stay.

This is a loose translation of a popular song made famous by Celia Cruz, who sang these words and brought tears to the eyes of those who fled.

Flights to Cuba were very unique and special. Many crew members did not understand the underlying reasons for these flights and treated them like ordinary ones. They were not! Everyone checked one or more suitcases. In addition, the 2 items, personal and carry-on, were always brought on and filled the overhead bins and the spaces under the seats. Things often stuck out a little further than they should, but caringly, we tried to help these folks.

In these carryons, were soap, shampoo, personal hygiene items etc. These were things that we, as a crew left in the hotel room the night before because they were of little or no value to us. We could go a the Dollar Store and get most of these things in large sizes for a buck. To those going to Cuba, it was gold! Once this was explained, it made the trip take a different feel. Many of these people were visiting grandparents whom they had never met. Grandchildren were meeting grandparents and family members for the first time.

On these short 45 - 50 minute flights, the service was simple. They ordered a Coke or orange juice, and that's it. The Americans ordered several mixed drinks and thought we had a bar on board.

You cannot go down to Cuba for vacation. There are approximately 13 things on a list that you could petition for a visa. A pilgrimage to a family's burial site, a religious mission, an educational project etc. It was not hard to get permission, but it was not like going to Jamaica, where you bought a ticket and a hotel room. Most families bring desperately needed items to family members.

As a crew, we could not spend the night there. In Cuba, there are no big hotel chains and suitable places for us to stay safely. We could not stay in a Bed and Breakfast in somebody's house or a boarding house. After an hour of refueling and cleaning the plane, we were back in Florida.

As the plane taxied in, I would take the useful, welcoming announcements in English and Spanish and then sing that song. People love it. They began to sing along and to cry. It was a very moving experience, and I loved each time we went. We also were able to get Cuban rum and cigars to bring back. We were not allowed off the plane, but the Captain would usually make a run to the duty-free shop.

On one trip, as I stood at the door with the captain to bid farewell, an older woman came up to us, grabbed our hands and began kissing them. She kept saying, "Gracias! Gracias!" Quite a moving experience.

I am so glad and honored that I could take part in this. My Spanish skills were a plus because I was often selected to work these flights. All Cuba flights had to have a certified Spanish speaker on board. This was not only for an explanation of the menu and little things but due for safety. If there was an emergency, instructions had to be given clearly in English and Spanish.

This is one of the many joys in life. I'm glad I was able to do it and teach others the joy that they were going to feel.

David Pinkwasser

It's Thanksgiving What's That Aroma?

What aromas come to mind on Thanksgiving? Do you think of turkey roasting in the oven, candied sweet potatoes, green bean casserole, hot biscuits browned to perfection etc.? Today was totally different. You could not make up this story if you tried! It's a scenario you wouldn't believe. Here's how it goes.

Today is early morning on Thanksgiving. This is a very happy day for all. People are traveling to see family and friends with anticipation of a beautiful meal and lively conversation. Faces that you haven't seen in a while and stories to be shared are being thought of in your mind. It is especially good for flight attendants. We get paid double time on Thanksgiving. If you are senior enough to get a trip that pays well, starts early and gets you home in time for dinner, you've caught the gold ring on the merry-go-round. (To the younger people who have no idea about the gold ring on a merry-go-round, ask an older person to explain it to you.)

We are now in the middle of a cross-country trip. There is a person in the front, lav who has been in there for about 20 minutes. I knock on the door and ask if everything is all right. The door swings open and a woman is in there with a small dog and is dipping it in the sink, which is filled with water and disintegrated paper towels. There is dog poop flowing in it, and she is trying to wash the dog. She starts going back to her seat, and I see that the tray and the walls all around her are covered in poop, both fresh and dried. The smell is horrific. The bathroom is, likewise, horrific. The woman tells me that she will take care of it. I tell her,

"YOU HAVE ALREADY DONE ENOUGH! I'LL TAKE CARE OF IT!!!!!"

I clean up her seating area with paper towels and disinfectant. I was now ready to work on the sink.

The water won't go down because the paper is clogging up the drain. I am wearing gloves and ask the other flight attendant to hold open a garbage bag while I pull pieces of paper and poop out of the drain and sink. The flight attendant lasts about 2 minutes and tells me that she can't do it. The smell is making her sick, and she runs away. I am now on my own. As I slowly remove the necessary items from the sink, the phone rings from the cockpit. They want to know what that smell is. They say, "Oh! I guess that lav will be inoperative for the rest of the trip, and we were about to use it. Too bad." I reply that I am fixing it. There are too many hours left on the trip to have only 1 bathroom. If you give me 10 more minutes, it will be ready.

All is now fresh and clean. I used deodorizers the best that I could, but there was still a slight stench in the front galley. It turns out that it was on my shoes, and I was tracking it. I cleaned the shoes and the floor, and now, believe it or not, I'm hungry, but the front galley is not the right place to eat.

I get to the back galley and am ready to eat, but I can still smell the aroma of dog poop. This is actually something that is in my head more than the actuality. The third flight attendant, who was not part of the clean-up, could not smell it. I and the second one could. We then had a plan. The third flight attendant had essential oils of lavender in her bag. She took a drop and rubbed it under both of our noses. Voila! Smell gone. This is also a trick that I use when I am for traumatic cases on the Critical Response Team for people who keep smelling odors that aren't there, i.e., burning rubber, jet fuel etc. We are all good now and ready to take this puppy home!

We are now at the Thanksgiving table, and the family asks how my day went. I told them that it was quite eventful, but not table talk. After dinner, I would be happy to share my day's experience with them.

David Pinkwasser

When you're a flight attendant and dealing with the public, you never know what direction you will go. That's part of the excitement of being a flight attendant. It is not a boring job. If you are flexible and can go with the flow, you'll love it.

*THE ADVENTURES OF THE RABBI
WHO BECOMES A FLIGHT ATTENDANT*

Can You Synchronize This?

When you think of synchronicity, swimming, gymnastics, acrobatics, marching etc., come to mind. This is a short tale of something you never thought of in your life: synchronized vomiting.

We are about 1 or 2 minutes from touchdown. The cabin is secure. There are 3 unaccompanied minors sitting in the 2nd row. They are referred to as UM's. As we are descending, the 2 of us are strapped in, and we notice that the first UM nearest the window begins to throw up. Then the second, and finally the third. To make matters worse, the adults on the other side smell the vomit, and the first adult throws up, then the second and finally the entire 2nd row. All 6 seats have people throwing up and not using the bags provided. There was nothing that we could do at that point, so we stayed seated, did the arrival PA, and asked people to stay seated when we arrived due to an accident up front. NO

ONE LISTENS! As soon as the door opened, the operation agent smelled the odor and quickly walked away from the plane. The passengers charged off the plane as quickly as possible to get away from the terrible smell.

We tried to clean up the kids the best we could, but the vomit permeated their clothing and shoes.

We brought the kids to the terminal. There were 2 out of 3 parents there. We explained what had happened, and they understood. The 3rd had to wait with the agent until the parents came. That was not pleasant, neither for the agent nor for the people checking in for the flight. (By the way, parents are supposed to be at the gate 45 minutes PRIOR to arrival). So much for that!

We cleaned what we could, but professional cleaners had to come in and finish the job. We ended up taking a 1-hour delay. The new group of passengers were not happy about that. When we explained the circumstances, it assuaged their anger, and they were glad to be in a clean environment.

Every day is an adventure. That's the beauty of this job: always something unexpected. Flexibility is the key.

David Pinkwasser

Is There A Doctor In The House?

IS THERE A DOCTOR IN THE HOUSE? These are words that you do not want to hear when flying. Those of us who are flight attendants have some medical training. We can handle most of the minor problems on board: nausea, cuts and bruises, headaches, falls, etc. We have trained to recognize more involved cases and, for the most part, are able to deal with them. HOWEVER, there are times when people appear to be having coronary problems, seizures, labor etc., when professional medical assistance would be indicated. Needless to say if there is no medical assistance on board, we would attempt to deal with the issues, and we are all trained in using the AED (Defribulator), if necessary, but professional assistance would be advisable.

THE ADVENTURES OF THE RABBI WHO BECOMES A FLIGHT ATTENDANT

Story 1

This was the worst medical emergency I had ever experienced. We had a heart attack going on in the front galley and a seizure toward the back of the aircraft. The chances statistically of this happening together on the same flight are pretty slim, but this is life. It did happen.

Two flight attendants were caring for the heart attack person in the front. I attended to the seizure in the rear. The person seizing had been doing so for about 10 minutes. I knew what to do and took care of the situation.

While this was happening, a call light came on, so I rushed to see what the problem was. The person said, "I know that you are very busy, but I did order a diet Coke about 15 minutes ago and I'm very thirsty." I responded that I was very sorry, but at this juncture, we are trying to save 2 people's lives. If we are able to get you a drink, we will. If not, I'm very sorry.

Fortunately, we had a deadhead crew on board. I asked one of them I they could go to my galley and get the "jackass" in row 15 a Diet Coke. Of course, they said sure, and of course, they rolled their eyes in amazement. Some people don't have any common sense or empathy for others.

With luck, my seizure passenger was all right when we landed. I did have to call the paramedics, as is standard procedure. The woman denied having a seizure, even though 20 people witnessed it. The paramedics asked me what the symptoms were, and I did a quick demo of what she was doing. They responded by saying, "Oh yes, that's a seizure." It is often common for people who have seizures to not remember the incident.

As for the cardiac passenger, he was still alive but not looking so good. He was taken to the hospital, and we have no idea what the outcome was. We are never informed of the results. We hoped for the best. We were able to keep him somewhat stable.

This was an incredible day. After the crises that we experienced, we each got a call from the CISM team that evening to inquire how we were doing and if any of us were shaken up by it. The deadhead crew also received a call.

Incidents like this can happen in a flash. You never know when it will occur and you have to always be prepared.

Story 2

We are flying from the east coast to Detroit. Somewhere near Columbus, I hear groaning coming from the front lav. These were not groans of excitement, such as a mile-high club member (these always take place in the back, not in the front where the whole plane can see 2 people going in together). They were sounds of pain. I knocked on the door to ask if they were all right. I only heard groans. I then announced: FLIGHT ATTENDANT COMING IN.

We all knew how to open the lav doors with a stir stick, but it wasn't working. The lock wouldn't move, and the stick kept breaking. After a few unsuccessful attempts, I used my house key. I was stronger, but I still wouldn't move the lock. I then deduced that the person was lying on the floor and leaning on the door. I pushed the door in and then the lock slid easily. I pulled the passenger out and brought her into the front galley.

The person was still moving and groaning, but then it stopped. I was about to begin CPR when the person now regained consciousness. I looked up, thanked God, and began to talk to the person. After 15-30 seconds, she passed out again. Checking her pulse and breathing, I was ready for my next plan as she then came back to consciousness. We were able to keep her awake, did whatever medical movements were necessary, and asked the captain to make an emergency landing. UNFORTUNATELY, the weather beneath us in

Columbus was snowy and windy. It was too dangerous to land there. We continued to Cleveland and were able to make a safe landing there. The paramedics came on the plane and took the passenger off. She informed us that she was actually going to Detroit. We informed her that was not going to happen today. She was going to the hospital in Cleveland to be checked out.

We learned that she had heart problems and did not take her meds while on vacation. This was not a smart move, and this heart episode was the result.

It is always important to take your medication as prescribed, even on vacation. You should also never pack it in checked luggage, but keep it with you on board in case you might need it. These are lessons to be learned by all.

Story 3

This brings us to the next incident. We were flying from the west coast to the east coast. The flight is now about an hour and fifteen minutes from landing. A passenger rings the call button and is having heart problems. We, of course, question the passenger and ask if she has had these problems before. The reply was that she had microvalve prolapse problems and had medication to control it. We asked when the last time she took her medication. She responded, "Two weeks ago before I went on vacation." Then we asked if she had the meds. She responded that it was home. We also asked when the last time she ate. She responded, "Last night. I didn't have any breakfast today. I'll eat when I get home." These were not the correct answers. The captain was informed and asked for special clearance to get through the heavy East Coast traffic. We made the one hour and fifteen-minute trip in thirty minutes. Pretty slick! She was the first one off the plane and we wished her well.

These are all reminders to us that it is important to take care of your health ALWAYS!!!!. This is not something that you forget if you're on vacation or taking a flight. Your body doesn't know this. You would think that this is very basic and that no one needs to be reminded of it, but as you can see, it is necessary. Good health is something you need to work on every day, especially if you are leaving your home surroundings. Please take it seriously and guard what is your precious gift.

David Pinkwasser

Serious Stuff—CISM-CoHeart-Lodo

CISM

Critical Incident Stress Management is a system of working with people who have had serious incidents that caused either acute or post-traumatic stress. It was developed in the 1980s by Jeffrey Mitchell and George Everly. It is a systematic way of reliving the trauma with a standard protocol that all practitioners of CISM are trained in.

After 9/11, most airlines realized that their CISM group needed to be expanded so that if an event, such as the attack on the World Trade Center, happens that affects many people, you need more numbers on your team.

I came on the team at that time. I felt that with my clergy and chaplaincy background, as well as my teaching career, I would be a good member of the team.

Each member of the team had to go through 2 levels of training, both basic and advanced. The basic covered 1 traumatic event. The advance covered 2 events that were occurring to the individual at the same time.

I was on call once a year. I carried a phone with me 24 hours a day. The time on duty was 15 days, so 2 people could cover each month.

Most of the calls came from scheduling or supervisors who were telling me about incidents that were happening that day, which might require a call from me. I also could receive calls from individuals themselves. This was rare, but each flight attendant and pilot had our 24 hr number with them at all times.

Some of the incidents were a death on board, a serious medical emergency beyond the normal ones, severe turbulence causing injuries, a decompression requiring oxygen masks, attacks on overnights by outsiders etc.

I got an average of about 5-10 calls a day. Every so often, I did not get any. That was rare. I always double checked to see if the phone was turned on or that the battery was still charged.

Here is an example of a typical call:

A flight attendant called me and told me that her leg was broken in 4 places. However, that was not her problem. The incident occurred during turbulence in the back galley. She was going through therapy, and all would heal. Her problem was that she could not sleep. When her eyes were closed, she smelled jet fuel and burning rubber. It kept her awake. Obviously, this was not happening. She was home, but in her mind, she was reliving the incident. She called the mental health agency from the airlines and was told that this was not a mental health issue. (I would tend to question that!). She was desperate and felt that this was her last possibility. I informed her that she called the right place. I had immediate credibility because I did the same job as her and knew exactly what she went through. This is what I told her.

I asked if she had eaten recently. She responded that she was not very hungry and was nibbling on cheese puffs. She also drank a bottle of wine every day.

My response was that she needed to ditch the cheese puffs and wine. I explained that you can't get better mentally if you are malnourished physically. She needed to drink a high-protein drink like Ensure, Boost or Slim Fast. It was important to get stronger. Secondly, alcohol is a depressant. These drinks would provide quenching of thirst and proper sustenance. If she needed to munch on something, ask someone to get you a barbecued chicken. You could pick pieces off when you are watching TV and get nourishment instead of empty calories.

The next step was sleeping. The room needed to be darkened. No TV or radio, unless it was environmental sounds or some type of elevator music that you would not recognize and want to hum along. Silence was the best. No reading or computers. Quiet and dark. Then, the bed should be sprayed with lavender mist. All of us carried this. Many of the hotels give this to us, and we all have it in our luggage. Under the nose, lavender essential oils would be the best. Place it on your upper lip under your nose. This should get the memories of the jet fuel and burning rubber to subside. If none was available, a menthol chest rub like Ben Gay would do the trick. I then wished her a good night.

The next day, I received a call from her. This was the best night's sleep she has had in several weeks, and she is feeling hopeful and energetic. I couldn't be thanked enough. I told her that her results were my thank you. I was delighted that it worked.

This is just a snippet of what the CISM Team does. I was honored to be part of it for 15 years and had many successes.

Co Heart Program

This was a wonderful program that lasted for about 5 years. It was curtailed, not because it wasn't successful, but due to budgetary restraints. I was a CoHeart Mentor. I flew with the brand-new hires on their first trip. I informed them that all of the stuff they learned in class was the book stuff. Now, we are in the real world. I know that the FAA and Safety Issues are important, but you also have to know how to crank out the drinks in a timely manner, especially if it is a short flight. Likewise, on a long flight, you have to pace yourself and not always do what the book says. You are dealing with real people, and every situation is different. It was a blast to show them the ropes. They were so appreciative, and all of my students have become lifelong friends.

In addition to serving, I showed them how to fill out the paperwork at the hotel to get everyone a room and how to navigate the restaurant and bar.

There are definitely certain protocols that are followed and expected. The crew was invited to dinner. I had an expense account and bought everyone dinner and drinks. It was a great first night on the job and at the hotel. It is a little different checking into a hotel as a crew as it is with your family.

The next day, the new hire was on their own, with my supervision, and at the end of the day, I left. And on day 3, this flight attendant was a well-oiled machine!

Great program. Great results.

LODO—Language of Destination Origin

This is a worldwide program to certify flight attendants in the language of the country they are entering. Those who passed the test, and unfortunately, many native speakers did not, did the announcements in English and the language they were certified in. I did my certification in Spanish. All of the announcements were done bi-lingually. It was important for people to be certified because you were doing more than just serving food and drinks. You were also doing the emergency briefings. If there was an emergency, those who did not know how to speak a language fluently could cause injury or death to a passenger because they didn't know how to say something.

I have always enjoyed these flights. The passengers were always surprised to hear that I was not Latino and spoke both Spanish and English without an accent. I told them that good professors in college did that.

The LODO program took me to places like Cuba, Mexico City, San Juan, Costa Rica etc. The people on each of these flights were interesting and diverse. As a Spanish major, we were taught the history, literature and culture of each Spanish-speaking country. On flights like this, it came in very handy. Being a New Yorker, I learned Spanish in a Puerto Rican environment, as opposed to the

Mexican environment in Arizona. I knew much of the slang, so when a person asked for"

JUGO DE CHINA, I knew it was orange juice and not Chinese juice. One who is not familiar with some of the localisms would have an idea of what the person was asking for.

Again, another adventure and a new experience. I am truly blessed.

*THE ADVENTURES OF THE RABBI
WHO BECOMES A FLIGHT ATTENDANT*

Help! Help! I'm Stuck!!

Help! Help! I'm Stuck. These are the words that came loud and cleared out of the back laboratory. The female flight attendant who was in the back of the plane heard these words and asked what assistance she needed. "I'm stuck! I'm stuck", she uttered. The flight attendant who knew, as we are trained, to open the door and see what the urgency was.

When the door opened, a corpulent woman had sat down on the toilet and did not put the seat down. She was now sitting on the bowl. When she flushed, the suction from the air-generated toilet caused a bond, and now she is stuck to the commode.

We are now about 10 minutes from landing, and she can't get off. It was decided that the safest place to be was right where she was. This was TRULY better than a seatbelt!

When on the ground, personnel were called to help this woman with her problem. A technician came in, using an implement, probably a screwdriver, and let the air out. This immediately released her from her position. Was it embarrassing and humiliating? YOU BET! What could you do? She was now able to go home and not be harmed, other than her pride, from the situation.

Shortly after, she tried to sue the airline because they did not have any signs posted saying that the seat must be down when using the toilet facilities. Now I ask you, have you ever seen such a sign in a public restroom? Of course not. I have seen signs in Asia where people are not used to Western plumbing and show people squatting, with a big X over the picture and a second picture with a drawing of someone sitting. In the United States, no.

The lawsuit never went anywhere. I'm not sure what the compensation was for her trouble, but it probably wasn't much, maybe a nice letter and a box of chocolates. I cannot say.

I am not sure that this is a lesson for anyone to learn. Most of us, if not all of us, know what to do in a restroom. Think about it the next time you use the facilities on an airplane and see if it doesn't give you some pause.

David Pinkwasser

Funny and Silly Stuff

Not everything on the plane was serious. There were a lot of fun things that we did to entertain and cheer up the passengers. When passengers are in a good mood, we are, too. I would often say some funny things or sing a song. For example:

St. Patrick's Day

We had a very Irish flight attendant who was on the crew. She had red hair and green eyes. Just a beauty. Her name was Katy. Before we made the announcement to secure the cabin, I announced that I would like to sing a special song to Katy McPherson (not her real name) on this happy day. I sang When Irish Eyes Are Smiling, with an Irish accent. The people went crazy and started singing with me. It was a great way to end a wonderful St. Patrick's Day.

TSA Song

Right after 9/11, there were extensive searches of people in addition to the normal screening. Every third person was a "selectee". Flights were delayed, and people came to the plane with their shoes off, belts in hand and altogether disheveled. Passengers were unnerved. I came up with a song that was to the tune of: It Had To Be You.

> It had to be you
>
> It had to be you
>
> Of all those around, they finally found
>
> Somebody like you
>
> So nobody else could give me a thrill
>
> Like security guard Bill
>
> It had to be you
>
> Had to be you
>
> Wonderful you.

Now, the nervousness and stress became laughter.

I learned as a clergy that when things are tense, some humor or a joke would ease the tension. Let me tell you, it worked like a charm.

Emergency Briefing

When we do the demo of the oxygen and lifevests, most people pick up a book or turn on their electronic devices and don't listen. I remedied this by saying:

Ladies and Gentlemen, if I could have your attention for just a moment as my wife and my ex-wife point out the safety features of this aircraft.

Immediately, you get attention.

To keep them interested, I might say:

In the event that this flight turns into a cruise, a lifevest is located under your seat.

Then, they would listen more to see if I was going to say anything else that was outrageous.

Sometimes I did, and sometimes not. Always keep them guessing.

Other times, I would say that I found a brown wallet on the floor. Does it belong to anyone? Now everyone is looking. Then I say:

Oh good, now that I have your attention, I'd like to point out the safety features of this aircraft.

Birthdays and Anniversaries

If it was a significant birthday or anniversary for someone. We would make a cake out of toilet paper and use stir sticks as candles. If it were a kid, we might make a crown out of peanuts and stir sticks. Then we would ask for the shades to be pulled down, the lights turned off, and all the call lights turned on. This looked like little candles. Everyone sang Happy Birthday. It was fun for all.

David Pinkwasser

Epilogue

Being both a rabbi and a flight attendant was a great experience. As a rabbi, I experienced the highest and lowest events in people's lives. I got to experience the joys as well as being a guide through the tough times.

As a flight attendant, I had the satisfaction of knowing that I could protect the plane and its passengers with my training. The knowledge of how to defend myself, put out a fire, cram luggage into spaces too small to fit, memorize all the FAA regulations, tend to the infirm and cheer people who were going to sad places. It was rewarding.

The question was asked of me, which job was more fulfilling. Obviously, being a rabbi was a calling. Definitely more meaningful. HOWEVER, as a flight attendant, I was more than your waiter. I did much more than just serve drinks and food. The variety of people that I encountered was life-changing. The various cities that we served gave us different clientele which made the experiences always challenging. The more you knew, the more equipped you were to meet these challenges. You knew what to expect. The bottom line is that I loved it all.

The experience of flying was a great way to end a very rewarding and meaningful career. It all added up to who I am today, and I like who I am.

I wish for people to follow their dreams and aspirations. It can be done. It may not happen when or where you plan, but don't give up if you see an obstacle, try to surmount it if you can't, and go in a different direction toward a new or altered goal. If you are persistent, it will happen. I know. I did it.

Made in the USA
Monee, IL
18 May 2024

58508328R00079